Kexin Yao

Complement of Special relativity and Limitation of General Relativity

AF 138604

Kexin Yao

Complement of Special relativity and Limitation of General Relativity

LAP LAMBERT Academic Publishing

Impressum / Imprint

Bibliografische Information der Deutschen Nationalbibliothek: Die Deutsche Nationalbibliothek verzeichnet diese Publikation in der Deutschen Nationalbibliografie; detaillierte bibliografische Daten sind im Internet über http://dnb.d-nb.de abrufbar.
Alle in diesem Buch genannten Marken und Produktnamen unterliegen warenzeichen-, marken- oder patentrechtlichem Schutz bzw. sind Warenzeichen oder eingetragene Warenzeichen der jeweiligen Inhaber. Die Wiedergabe von Marken, Produktnamen, Gebrauchsnamen, Handelsnamen, Warenbezeichnungen u.s.w. in diesem Werk berechtigt auch ohne besondere Kennzeichnung nicht zu der Annahme, dass solche Namen im Sinne der Warenzeichen- und Markenschutzgesetzgebung als frei zu betrachten wären und daher von jedermann benutzt werden dürften.

Bibliographic information published by the Deutsche Nationalbibliothek: The Deutsche Nationalbibliothek lists this publication in the Deutsche Nationalbibliografie; detailed bibliographic data are available in the Internet at http://dnb.d-nb.de.
Any brand names and product names mentioned in this book are subject to trademark, brand or patent protection and are trademarks or registered trademarks of their respective holders. The use of brand names, product names, common names, trade names, product descriptions etc. even without a particular marking in this work is in no way to be construed to mean that such names may be regarded as unrestricted in respect of trademark and brand protection legislation and could thus be used by anyone.

Coverbild / Cover image: www.ingimage.com

Verlag / Publisher:
LAP LAMBERT Academic Publishing
ist ein Imprint der / is a trademark of
OmniScriptum GmbH & Co. KG
Heinrich-Böcking-Str. 6-8, 66121 Saarbrücken, Deutschland / Germany
Email: info@lap-publishing.com

Herstellung: siehe letzte Seite /
Printed at: see last page
ISBN: 978-3-659-58412-1

Kexin Yao

Complement of Special relativity and Limitation of General Relativity

Table of Contents

Complement of Special relativity and Limitation of General Relativity

Abstract: For a set of forces being in equilibrium, the equilibrium state will not vary from one observer to another. This generally acknowledged fact has been called as the Force Equilibrium Invariance Axiom. On the basis of this axiom, the force transformation formula can be derived when an object is in motion, so that the force, length, time and mass can be calculated as per the basic transformation formula for the special relativity. According to the results of experimental analysis obtained by J.C.Hafele and R.E.Keating, the concepts of absolute time delay and relative time delay have been put forward herein, so as to solve any problems on the traveling velocity and records of any object being in motion at the real-time. For the properties of the force, the force applied on an object being accelerated will be deemed as the energy transfer force, while the universal gravitation will be deemed as the applied force of field. Both forces have the different basic properties. It is unscientific to consider both forces being the same properties. Based on the Force Equilibrium Invariance Axiom, it is deduced that the gravitational mass can only be the constant irrelevant to the motion of an object. Therefore, the principle of equivalence will not be tenable in theory. Only if the motion velocity of an object is very small relative to the velocity of light, can the principle of equivalence be deemed being approximately tenable. Under this condition, the general relativity can be only consistent with the practical deducing principle. However, if the motion velocity of an object is relatively large, there will be a significant difference between the constant gravitational mass and the inertial mass; as a result, there may be a deviation between the inference on the general relativity and the practical deducing principle. Take the Black Hole as an example. Even if the Black Hole is made almost completely of neutrons, its actual volume is one million times greater than its theoretical volume. Therefore, it is concluded that the Black Hole cannot be deemed as the substance composed of real atoms. It also shows accordingly that, according to the general relativity of theoretical source of Black Hole, the analysis results obtained when an object is in motion at high-velocity can not conform to the reality.

Keywords: Force equilibrium; Force transformation; Relative motion; Circular motion; Relative time delay; Absolute time delay; Energy transfer force; Applied force of field; Gravitational mass; Inertial mass; Equivalence principle; Black Hole

Chapter 1   Introduction

There are four fundamental physical quantities in the theory of mechanics, i.e. length l, time t, mass m and force F. It must be pointed out according to the special relativity that, if the inertial system Z' is in motion at the velocity v relative to the inertial system Z, it will be observed in Z that, the length l of a stationary object in Z' will be translated into $l' = l\sqrt{1-v^2/c^2}$, the time interval $t(t_2 - t_1)$ into $t' = t\sqrt{1-v^2/c^2}$ and the mass m into $m' = m/\sqrt{1-v^2/c^2}$. However, it isn't specifically and clearly interpreted in the special relativity whether or how to transform the stationary force in Z'. Therefore, in order to achieve a comprehensive transformation of the fundamental physical quantities in the special relativity, it is necessary to analyze and explore the transformation of force.

Moreover, relative to the rest length, the rest mass is constant; the time is running forward along with the time variation, and always changing its running records; the running velocity of time is determined by the size of time interval $t(t_2 - t_1)$; if t is large, the time will run quickly; if t is small, the time will run slowly, i.e. the time delay. However, $t' = t\sqrt{1-v^2/c^2}$ is the only transformation formula for an object being in relative motion, where neither the real clock running conditions nor the real clock running records of the stationary object can be determined or reckoned. For example, when A and B are in relative motion, if the rest time interval of B is expressed as t, it is judged by B that B is in motion at v relative to A, the time interval of A will be expressed as $t' = t\sqrt{1-v^2/c^2}$; however, if the rest time interval of A is also expressed as t and it is also observed that B is in motion at $-v$ relative to A, it is judged by A that the time interval of B will be expressed as $t' = t\sqrt{1-v^2/c^2}$. Obviously, according to the contradictory judged results of A and B, we cannot determine the real-time running conditions of A and B only based on the time transformation formula. How to determine the real-time running records of an object?

It is judged by the general relativity that the force applied on an accelerating object and the universal gravitation will be of the same properties. However, from in-depth analysis, these two forces are fundamentally different in some respects; so, it is unscientific to suppose that both forces be identical. And, as a theoretical basis of general relativity, the equivalence principle is merely deemed as a supposition. Due to lack of the theoretical analysis and convincing experimental proof, the most obvious and fundamental problem is that, when the object is in motion at v, whether is the gravitational mass expressed as $m' = m\sqrt{1 - v^2/c^2}$ the same as the inertial mass? Obviously, it is critical for the tenable gravitational mass to determine whether the equivalence principle is correct as per this transformation formula. Therefore, it is necessary for general relativity researchers to analyze and solve this critical problem.

Chapter 2   Transformation of Force Complemented to the Special Relativity

2.1 Force Equilibrium Invariance Axiom and force transformation formula

After the object (1 kg) has been placed on a spring balance, the pointer of spring balance is set at the 1 kg scale position, which remains 1kg scale not only to someone being stationary relative to the spring balance but also to anyone being in motion relative to the spring balance. This fact shows that a set of forces being in equilibrium in a reference frame can be observed by any other observers in the reference frame. Or rather, a set of forces being in equilibrium will never vary from one observer to another in the reference frame. This is a fact accepted in people's daily life, called as the Force Equilibrium Invariance Axiom. The force transformation formula can be derived on the basis of this Force Equilibrium Invariance Axiom.

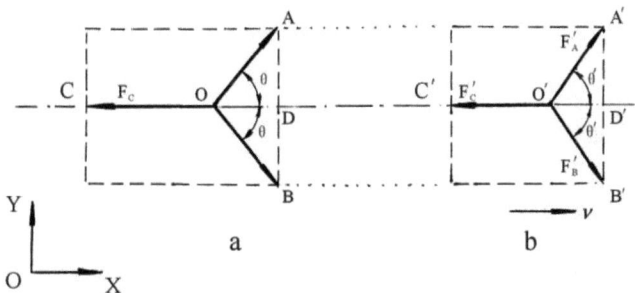

Fig. 1: Changes in Size of the Force being in Motion

The three forces (F_A , F_B and F_C) are in equilibrium (as shown in Fig. 1a), including: Applied force of electric field, universal gravitation or spring force. As shown in Fig. 1a, the length of OA, OB or OC represents the size of F_A , F_B or F_C , respectively; F_C is parallel to the X axis; F_A is equivalent to F_B ; the included angle between F_A or F_B and the X axis is θ (irrespective of the positive or negative angle of θ).

Since the three forces are in equilibrium, it also can be referred to as $F_A \cos\theta + F_B \cos\theta = F_C$. Since F_A is equivalent to F_B , for simple derivation, if $F_A = F_B = F$, it also can be referred to as $2F \cos\theta = F_C$.

The three forces (as shown in Fig. 1a) are in motion at v parallel to the X axis (as shown in Fig. 1b). According to the special relativity, the length along the v direction (X direction) will be shortened; F_C (as shown in Fig. 1b) will be shortened as F_C' ; F_A and F_B will be shortened as F_A' and F_B' respectively. According to the Force Equilibrium Invariance Axiom, F_C' , F_A' and F_B' are still in the state of equilibrium; F_A' is naturally equivalent to F_B' ; if $F' = F_A' = F_B'$, it also can be referred to as $2F' \cos\theta' = F_C'$.

The relation between F_C and F_C' can be inferred according to any changes in applied force of electric field being in motion.

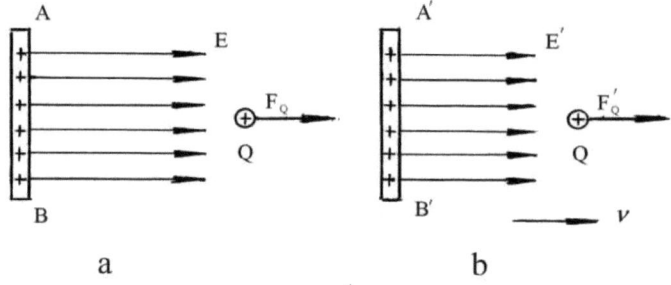

a   b

Fig. 2: Electric Field being in Motion Parallel to the Intensity Line of Electric Field E

The electric charge Q is in a uniform electric field generated by the "infinitely great" charged plate (as shown in Fig. 2a). E represents the intensity of electric field; AB represents a certain cross-section of this charged plate; obviously, the applied force of Q will be expressed as $F_Q = EQ$. As shown in Fig. 2b, after AB is in motion at the velocity v parallel to E, AB has been transformed as $A'B'$. According to the

8

special relativity, after it is in motion along the length vertical to the direction of v, the size of length will be invariable; it also can be referred to as $A'B' = AB$, naturally resulting in the intensity of electric field of $A'B' : E' = E$. Therefore, when $A'B'$ is in the electric field, the applied force of Q will be expressed as: $F_Q' = E'Q = EQ = F_Q$. It follows from this fact that, after the force F_Q is in motion along the direction parallel to F_Q, the size of force will be invariable ($F_Q' = F_Q$). For the force (as shown in Fig. 1), it also can be referred to as $F_C' = F_C$. Since $F_C' = 2F' \cos\theta'$ and $F_C = 2F \cos\theta$, it also can be referred to as $F' \cos\theta' = F \cos\theta$.

$$F' = F \frac{\cos\theta}{\cos\theta'}$$

As can be seen from Fig. 1:

$$\cos\theta' = \frac{O'D'}{\sqrt{O'D'^2 + A'D'^2}} \qquad \cos\theta = \frac{OD}{\sqrt{OD^2 + AD}}$$

On account of the same length being vertical to the direction of v ($A'D' = AD$), according to the special relativity, the length $O'D'$ along the direction of v can be calculated as per the length transformation formula ($O'D' = OD\sqrt{1 - v^2/c^2}$). According to the equation $A'D'$ and $O'D'$, the $\cos\theta'$ can be transformed as:

$$\cos\theta' = \frac{OD\sqrt{1 - v^2/c^2}}{\sqrt{OD^2(1 - v^2/c^2) + AD^2}}$$

$$F' = F \frac{\cos\theta}{\cos\theta'} = F \frac{OD}{\sqrt{OD^2 + AD^2}} \cdot \frac{\sqrt{OD^2(1 - v^2/c^2) + AD^2}}{OD\sqrt{1 - v^2/c^2}}$$

$$= F \sqrt{\frac{(OD^2 + AD^2) - OD^2 v^2/c^2}{(OD^2 + AD^2)(1 - v^2/c^2)}}$$

After $\cos\theta' = OD/\sqrt{OD^2 + AD^2}$ is substituted into the above equation, it also can be referred to as:

$$F' = F\sqrt{\frac{1-\cos^2\theta\, v^2/c^2}{1-v^2/c^2}} \tag{1}$$

Since $F' = F\cos\theta/\cos\theta'$ and $\cos\theta' = F\cos\theta/F'$, it also can be referred to as:

$$\cos\theta' = \cos\theta\sqrt{\frac{1-v^2/c^2}{1-\cos^2\theta\, v^2/c^2}}$$

$$\theta' = \cos^{-1}\cos\theta\sqrt{\frac{1-v^2/c^2}{1-\cos^2\theta\, v^2/c^2}} \tag{2}$$

The above equations (1) and (2) will be deemed as the transformation formula when the force is in motion at v. The force transformation formula is the same as the length, time and mass transformation formula, i.e. the fundamental transformation formula for special relativity.

2.2 Electric field distribution of charged particles being in motion as derived from the force transformation formula

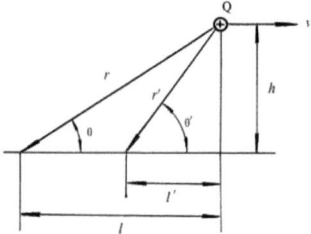

Fig. 3: r will be transformed as r′ when the charged particle is in motion at v.

The positively charged particle Q is in motion at v (as shown in Fig. 3). The rest length l (as shown in Fig. 3) will be shortened as $l' = l\sqrt{1-v^2/c^2}$ after motion. The size of height being vertical to the direction of v will be invariant before and after motion. r (as shown in Fig. 3) will be shortened as r′ after motion. It can be seen from Fig. 3 that if $l = r\cos\theta$, $l' = l\sqrt{1-v^2/c^2} = r\cos\sqrt{1-v^2/c^2}$ and $h = r\sin\theta$, it also can be referred to as:

$$r' = \sqrt{l'^2 + h^2} = \sqrt{r^2 \cos^2 \theta \left(1 - v^2/c^2\right) + r^2 \sin^2 \theta}$$

$$= r\sqrt{1 - \cos^2 \theta \, v^2/c^2} \tag{3}$$

Since $h = r\sin \theta = r'\sin \theta'$, it can be referred to as $\sin \theta' = \sin \theta \, r/r'$. After the equation (3) is substituted into the above equation, it also can be referred to as:

$$\sin \theta' = \sin \theta \Big/ \sqrt{1 - \cos^2 \theta \, v^2/c^2} \tag{4}$$

Since Q is the electric field (the intensity of electric field E), the applied force will be referred to as F = EQ, i.e. the intensity of electric field E is proportional to the applied force F; so, the force transformation formula also can be referred to as the transformation formula for the intensity of electric field as follows:

$$E' = E\sqrt{\frac{1 - \cos^2 \theta \, v^2/c^2}{1 - v^2/c^2}} \tag{5}$$

The equation (5) can be transformed as:

$$E' = E\sqrt{\frac{1 - \cos^2 \theta \, v^2/c^2}{1 - v^2/c^2} \cdot \frac{\left(1 - \cos^2 \theta \, v^2/c^2\right)\left(1 - v^2/c^2\right)}{\left(1 - \cos^2 \theta \, v^2/c^2\right)\left(1 - v^2/c^2\right)}} E$$

$$= \frac{\left(1 - \cos^2 \theta \, v^2/c^2\right)^{3/2}}{\left(1 - v^2/c^2\right)^{3/2}} \cdot \frac{1 - v^2/c^2}{1 - \cos^2 \theta \, v^2/c^2} E$$

$$= \frac{1 - v^2/c^2}{\left(\dfrac{1 - v^2/c^2}{1 - \cos^2 \theta \, v^2/c^2}\right)^{3/2}} \cdot \frac{E}{1 - \cos^2 \theta \, v^2/c^2}$$

Since $1 - v^2/c^2 = 1 - \left(\sin^2 \theta + \cos^2 \theta\right)v^2/c^2$, the above equation can be transformed as:

$$E' = \frac{1 - v^2/c^2}{\left(1 - \dfrac{\sin^2 \theta \, v^2/c^2}{1 - \cos^2 \theta \, v^2/c^2}\right)^{3/2}} \cdot \frac{E}{1 - \cos^2 \theta \, v^2/c^2}$$

After equations (3) $\sqrt{1 - \cos^2 \theta \, v^2/c^2} = r'/r$ and (4) $\sin \theta' = \sin \theta \Big/ \sqrt{1 - \cos^2 \theta \, v^2/c^2}$ are substituted into the above equation, it also can be referred to as:

$$E' = \frac{1 - v^2/c^2}{\left(1 - \sin^2 \theta' \, v^2/c^2\right)^{3/2}} \cdot \frac{r^2}{r'^2} E$$

Since $E = kQ/r^2$, after it is substituted into the above equation, it also can be referred to as:

$$E' = K \frac{1 - v^2/c^2}{\left(1 - \sin^2 \theta' v^2/c^2\right)^{3/2}} \cdot \frac{Q}{r'^2} \tag{6}$$

The equation (6) will be deemed as the electric field distribution formula for charged particle being in motion at v. By comparison, this equation will be identical to the formula derived by electrodynamics.

When the force-applied object is in motion, the force will be transformed like as the length, time and mass. For example, when the charged particle iss are in motion in a magnetic field, the Lorentz force applied on the particles will be transformed. See supplementary paper 1

Chapter 3   Absolute time delay and relative time delay complemented to the Special Relativity

3.1 Two types of object motion

In our daily life, we can observe two types of object motion. The first one is relative motion in which either side can observe the opposite side in motion. For example, we can see from the train window that an opposite train is traveling, which might be traveling or stopping, while our train is traveling as a matter of fact. This is relative motion. According to the special relativity, each side being in relative motion may consider that the time interval of the opposite side being in motion is shortened, i.e. the time delay. The second type of motion is circular motion, different from relative motion. For example, the ferris wheel cabin of playground is rotating around the center of ferris wheel, or the man-made satellite or the Moon is rotating around the Earth. Unlike the relative motion, the circular motion refers to a universally acknowledged motion in which the circular object can revolve around the center of a circle rather than the motion in which the circular object and the center of a circle can revolve around each other. Taking the ferris wheel as an example, the center of ferris wheel and the person on the stationary ground relative to the center of ferris wheel can consider that the ferris wheel cabin is rotating around the center of ferris wheel, while the person in ferris wheel cabin can also consider that he/she is rotating around the center of ferris wheel, i.e. the circular motion of ferris wheel cabin will be deemed as an accepted motion rather than a relative motion. Similarly, the motion of the man-made satellite or the Moon revolving around the Earth will be deemed as an accepted motion as well.

The accepted velocity of circular motion can be defined as the absolute velocity. Naturally, the velocity of relative motion can be defined as the relative velocity.

If the rotational velocity of an object being in circular motion is set as v, and the time interval of the center of rotation is set as T, then according to the special relativity, the time interval of the object being in circular motion will be inevitably expressed as:

$$t' = t\sqrt{1 - v^2/c^2} \tag{7}$$

Obviously, here t' represents the time interval between the object being in circular motion and the center of circle. Such t' is not relative, but absolute. Relative to t, t' is reduced. A reduced t' indicates that the time motion velocity is low, i.e. time delay. Such accepted time delay t' will be defined as absolute time delay, e.g. time delay of the Earth relative to the Sun, time delay of the Moon relative to the Earth, etc.

Next, when it comes to the time motion condition, we will only use the term "time interval"; when the time interval is relatively reduced, it will be referred to as time delay.

There are a variety of circular motions in the universe; each has its own different absolute velocity and absolute time delay. How to compare with their sizes of motion? First of all, there must be a standard to make comparison. For us, of course, the time interval t of the rotation center of the Earth (i.e. the Earth's south and north poles, say 1 hour) will prevail.

Relative to t, the time interval t_m of the Moon is reduced, i.e. t_m in relation to t is time delay. Assuming that the Moon moves around the Earth at a velocity of v_m, then the time interval of the Moon will be expressed as:

14

$$t_m = t\sqrt{1 - v_m^2/c^2} \tag{8}$$

If the Earth moves around the Sun at v, and the time interval of the Sun is expressed as t_0, then it can be derived that $t = t_0\sqrt{1 - v^2/c^2}$, namely,

$$t_0 = t/\sqrt{1 - v^2/c^2} \tag{9}$$

Obviously, t_0 is greater than t, so t_0 cannot be referred to as time delay. If a certain planet of the Sun moves around it at v_x, the time interval of x planet will be expressed as:

$$t_x = t_0\sqrt{1 - v_x^2/c^2} = t\sqrt{1 - v_x^2/c^2}/\sqrt{1 - v^2/c^2} \tag{10}$$

It can be observed that, if $v_x > v$, then $t_x < t$. Relative to t, t_x will be defined as time delay; on the contrary, the time speeds up.

If a certain satellite moves around the Moon at V_s, then the time delay of the Moon's satellite will be expressed as:

$$t_s = t_m\sqrt{1 - v_s^2/c^2} = t\sqrt{1 - v_s^2/c^2}\sqrt{1 - v^2/c^2} \tag{11}$$

The comparison between time intervals of other objects being in circular motion can be analogized according to the above calculation methods.

Each celestial body in the universe has its own different time interval, and each generally rotates by itself. Relative to the axis of rotation, the celestial body has its different velocity of rotation at different locations; there is also a difference between time intervals. In the case of the Earth, the equatorial radius of the Earth is about 6,378km; it takes 24 hours for the Earth to rotate around its axis once; the velocity of rotation of the Earth's land surface can be calculated as 0.464km/s. Along with an increase in terrestrial latitude, the velocity of rotation of different points on the Earth's land surface will decrease. If the terrestrial latitude somewhere on the Earth is expressed as θ, then the velocity of rotation of this point relative to the axis here will be expressed as $v_\theta = 0.464\cos\theta\,km/s$ and the time interval at this point will be expressed as:

$$t_\theta = t\sqrt{1 - 0.2153\cos^2\theta\big/c^2}\qquad\qquad(12)$$

It is concluded from the above analysis that any celestial body or its different location has its own fixed time interval. The locations with the same time interval can be divided into equal time interval zones. For example, the Earth's surface with the same latitude can be divided into equal time interval zones. If a certain object is placed at t_1 time zone, the time motion velocity of such object will operate as per t_1; if such object is placed at t_2 time zone, its time motion velocity will operate as per t_1, and so on.

3.2 Experiment conducted by J·C·Hafele and R·E·Keating

In 1971, J·C·Hafele and R·E·Keating carried out an experiment for the relationship between time delay and motion velocity. They placed four caesium atomic clocks on an aircraft stopping near the equator. After the aircraft traveled along the equator from east to west so as to make a complete cycle around the Earth, it was found that the average reading of the four caesium atomic clocks was 273 × 10^{-9} seconds faster than that of the caesium atomic clock placed on the ground (surface phenomenon was negative time delay). However, after the aircraft traveled along the equator from west to east so as to make a complete cycle around the Earth, it was found that the average reading of the four caesium atomic clocks was 59 × 10^{-9} seconds slower than that of the caesium atomic clock placed on the ground (Reference 1). Why could such a result occur? As previously described, the absolute

16

time delay of an object or any point on the ground relative to the Earth's axis will depend on its motion velocity relative to the Earth's axis. When both the aircraft and equatorial ground are rotating around the Earth's axis, the velocity of rotation of equatorial ground around the Earth's axis will be set as v_1; the flight velocity of the aircraft will be set as v. When the aircraft is flying to the west, the motion direction of the aircraft will be opposite to the direction of the Earth's rotation and the actual velocity of rotation of the aircraft around the Earth's axis will be set as $v_2 = v_1 - v$. When the aircraft is flying to the east, the motion direction of the aircraft is consistent with the direction of the Earth's rotation and the actual velocity of rotation of the aircraft around the Earth's axis will be set as $v_3 = v_1 + v$. It can thus be seen that the time interval of equatorial ground relative to the Earth's axis (time interval T) can be expressed as $t_1 = t\sqrt{1 - v_1^2/c^2}$; the time interval of aircraft to the west relative to the Earth's axis can be expressed as $t_2 = t\sqrt{1 - (v_1 - v)^2/c^2}$; and the time interval of the aircraft to the east relative to the Earth's axis can be expressed as $t_3 = t\sqrt{1 - (v_1 + v)^2/c^2}$. Obviously, if $t_2 > t_1$, namely, the time motion velocity of the aircraft to the west is faster in relation to the Earth's axis, the absolute time delay of the aircraft to the west is less than the absolute time delay of equatorial ground (t_2 is large; time delay is small); if $t_3 < t_1$, the absolute time delay of the aircraft to the east is greater than the absolute time delay of the equatorial ground. J·C·Hafele and R·E·Keating drew the following conclusions by calculations on the basis of such difference: The caesium atomic clock on the aircraft to the west would be 275×10^{-9} faster than the caesium atomic clock on equatorial ground, consistent with the measured readings. In addition, they reckoned from $t_3 < t_1$ that the caesium atomic clock on the aircraft to the east would be 40×10^{-9} slower than the caesium atomic clock on the equatorial ground. Outwardly, although there is a major difference between this result and the measured readings (59×10^{-9} seconds), this is only a comparison difference, rather than a calculation error. For example, if the height of a wall is up to 300cm and the height of a tree is up to 301cm, then the tree is 1cm higher than the wall. However, if the calculated height of the tree is up to 304cm, then the tree is 4cm higher than the wall according to calculations. Obviously, there is a major difference between 4 and 1; however, this difference can be deemed as a comparison difference rather than a calculation error. The calculation error will be deemed as the error between 304 and 301, i.e. the error is less than one percent (1%).

In accordance with our requirements, one hour of the equatorial caesium atomic clock, i.e. 3,600 seconds, is set as the standard time interval t_1. When the equatorial caesium atomic clock has gone forward by 50 t_1 (i.e. 1.8×10^5 seconds) after the aircraft completes one circle around the Earth, the measured motion time of the caesium atomic clock on the aircraft to the east will be $50t_3=1.8\times10^5\text{s}-59\times10^{-9}\text{s}$, but the theoretically calculated motion time of the caesium atomic clock is $50t_3'=1.8\times10^5\text{s}-40\times10^{-9}\text{s}$. The ratio between the calculated value and the measured value (t_3'/t_3) can be expressed as $\left(1.8\times10^5 - 40\times10^{-9}\right)/\left(1.8\times10^5 - 59\times10^{-9}\right)$. Thus it can be seen that there is a relatively small error between the theoretically calculated value t_3' and the measured value. Therefore, we may consider that the analysis results are consistent with the experimental results obtained by J·C·Hafele and R·E·Keating.

The experimental results obtained by J·C·Hafele and R·E·Keating have proved that the velocity of rotation of the equatorial ground, the aircraft to the west or the aircraft to the east around the Earth's axis could be referred to as absolute velocity; They also demonstrated that the circular motion velocity could be referred to as absolute velocity andthe circular motion time delay as absolute time delay. Moreover, it shows the appropriateness of the method for determining the time interval as per the division of time zones.

3.3 Relative time delay and absolute time delay & predication of real-time progress

In the experiments conducted by J · C · Hafele and R · E · Keating, the flight velocity of the aircraft flying to the east is the same as to the west, and people on the equatorial ground are supposed to observe that the time delay results of the aircraft to the east can be identical with those of the aircraft flying to the west. However, the experimental results have shown that the time delay results of the aircraft flying to the east are different from those of the aircraft to the west; especially the time of the aircraft to the west, instead of being delayed, speeds up. It is proved beyond doubt that the relative time delay judged by the people as per the relative velocity is fundamentally different from the real absolute time delay of a moving object. The experimental results obtained by J · C · Hafele and R · E · Keating affirmed the correctness of the method for calculating the absolute time delay as per the circular motion velocity. Moreover, the muon lifetime experiments have also confirmed that the relative time delay could be deemed as a real observation. How to understand the two seemingly different but truthful judgments?

In connection with any changes in the length of an object, we may understand the above-mentioned two different judgments. As we know, when an object is in motion at a velocity v, the length l of the object along the direction of motion will be shortened as $l' = l/\sqrt{1 - v^2/c^2}$. We observe that l' of the object being in motion is real, but the length l of the object is also real without any changes, i.e. both l and l' are real. Observers using different reference systems can obtain such an observation. Changes in time are similar to changes in length. That there is distinction between relative time delay and absolute time delay with respect to time could also be observed by observers using different reference systems.

Changes in length and time are related to the motion velocity observed by the observer. The observer on the equatorial ground being in motion at a velocity v_1 observes that the relative velocity of the aircraft flying to the west or the aircraft to the east relative to him can be expressed as v; both the length contraction and time delay of the two aircrafts are naturally identical. However, an observer not standing on the equatorial ground will not observe that the relative velocity of the aircraft flying to the west is identical with the relative velocity of the aircraft flying to the east relative to him. For example, an observer on the Earth's north or south pole observes that the relative velocity of the aircraft flying to the west relative to him is expressed as $v_1 - v$ while the relative velocity of the aircraft flying to the east relative to him is expressed as $v_1 + v$; both the length contraction and time delay of the two aircrafts are significantly different. That is because observers using different reference systems will observe that the same object will be provided with different length contractions and relative time delay, which, though different, will be deemed as the real observation.

Under normal circumstances, the relative time delay is different from the absolute time delay. However, if an observer is located at the center of a circle of the object being in circular motion or within the stationary area relative to this center of a circle, i.e. the area where the absolute motion of the object has been accepted as mentioned before, then the observer will observe the relative velocity of the object being in circular motion relative to him, i.e. the absolute velocity of the object being in circular motion, and the relative time delay of the object will be referred to as the absolute time delay. For example, the relative time delay of the particle being in circular motion in synchrocyclotron as observed by us will be referred to as the real absolute time delay of the particle.

The experimental results obtained by J · C · Hafele and R · E · Keating show that when a stationary object undergoes a motion process before returning back to rest, the rest length and rest mass of the object will not change, but the time progress of the moving object is different from the time progress of a non-moving object. (When the aircraft returns to the ground, the time progress of the aircraft's caesium atomic clock is different from the time progress of the ground caesium atomic clock). Also, such changes in time progress will be related to the motion history of an object, e.g. the aircraft's flight direction, velocity, flight time, etc.

Based on the foregoing analyses, we can easily calculate the corresponding time process with respect to different motion histories of an object. For example, according to the time standard of one hour (as expressed in h) relative to the Earth's axis, we can leave a clock at the terrestrial latitude (45°) on the Earth for n_1h, then on the Moon for n_2h and then on the Sun's x planet for n_3h, and then return it to the ground. According to the above motion history, we can calculate the cumulative absolute time process of this clock as per the following formula:

$$\left(n_1\sqrt{1-0.2153/2c^2}+n_2\sqrt{1-v_m^2/c^2}+n_3\sqrt{1-v_x^2/c^2}\Big/\sqrt{1-v^2/c^2}\right)h$$

We can give a very simple and clear explanation of the twin paradox problem. For example, twins A and B can be placed at two locations with an absolute time interval of t_A and t_B; if t_A is less than t_B, i.e. the time motion velocity of time zone t_A is slow, when A and B meet each other at one location after a period of time, A will inevitably become younger than B; on the contrary, if t_A is greater than t_B, B will inevitably become younger than A. Take the experiment conducted by J · C · Hafele and R · E · Keating as an example. Suppose B is located on the equatorial ground, when A travels by aircraft to the east around the Earth and then returns to the ground to meet B after a period of time, A will inevitably become younger than B; however, when A travels by aircraft to the west around the Earth and then returns to the ground to meet B after a period of time, B will inevitably become younger than A. Please see supplementary paper 2 as reference.

Chapter 4   Applicable Scope and Limitation of General Relativity

4.1 Problems on general relativity

According to the general relativity, the force applied on an object being accelerated will be equivalent to the universal gravitation. However, as a result of analysis, the force applied on an object being accelerated will be deemed as the energy transfer applied force. For example, Object A can transfer its energy to Object B (mass m). B moves for a distance S under the acting force F of A at the time t; the output energy of A will be expressed as FS. Since B moves under the acting force F at the acceleration of a , if $S = at^2/2$, it will be also referred to as $FS = Fat^2/2$. B has obtained the velocity $v = at$ under the acting force F at the time t; the energy obtained by B will be also referred to as $mv^2/2 = ma^2 t^2/2$. The output energy of A will be equivalent to the input energy of B; so the following formula can be obtained:

$$\frac{Fat^2}{2} = \frac{ma^2t^2}{2}$$

This equation will be deemed as the formulation of accelerating acting force: $F = ma$; that is to say, $F = ma$ refers to the applied force of A for which B can obtain the energy. As can be seen that as long as there is the energy transfer force to exist, it will be accompanied by energy transfer process. The universal gravitation will be deemed as the applied force of field. There is no any relationship between the applied force of field and the energy transmission. Even if an object being in motion can obtain the kinetic energy under the universal gravitation, this kinetic energy can be resulted from the potential energy conversion of the object, rather than the gravitational output energy.

In addition, the universal gravitation will be deemed as the applied force of field.

The field will be mainly characterized by the constant field flux of any surface, i.e. there is no any correlation between the field flux and the motion of the observer, while the size of m of energy transfer force $F = ma$ is related to the motion velocity, i.e. $m' = m/\sqrt{1 - v^2/c^2}$. If the energy transfer force is equivalent to the universal gravitation, the observer being in motion at v can observe that m generating the gravitational field will be increased to $m' = m/\sqrt{1 - v^2/c^2}$, while the field flux of any one surface will be increased to $1/\sqrt{1 - v^2/c^2}$ accordingly. This judgment is clearly inconsistent with the basic properties of the field; it also means that the energy transfer force is not possibly equivalent to the universal gravitation.

Further, the energy transfer force is generally constant and the acceleration is also constant accordingly. However, the universal gravitation is generally inversely proportional to the square of the distance between gravitational objects; the acceleration generated by the object in the gravitational field is naturally inversely proportional to the square of the distance between gravitational objects; namely, the acceleration being in motion in the gravitational field will never be constant, and there is a fundamental difference between the energy transfer force and the universal gravitation.

Most importantly, the equivalence principle is the basis of general relativity. It is supposed according to the equivalence principle that the gravitational mass is equivalent to the inertial mass. If this supposition is tenable, the observer being in motion at a relatively high velocity v relative to the Sun will observe that the Sun and its planets are in motion at the velocity $-v$ (the rotating velocity v of planets around the Sun is relatively small and can be ignored); their gravitational mass can be increased $1/\sqrt{1-v^2/c^2}$ times. Thus, the universal gravitation between the Sun and its planets can be inevitably increased $1/(1-v^2/c^2)$ times. As a result, the universal gravitation of the Sun on its planets will be greater than the centrifugal force of the Sun's planets on the Sun; the planets are absorbed by the Sun; all the planets have disappeared; the solar system has become the lonely Sun. Obviously, the observer being in motion at v cannot see this result. Therefore, the gravitational mass is unlikely equal to the inertial mass; the equivalence principle will not be theoretically tenable; and there is naturally also a problem on the general relativity.

4.2 Gravitational mass of an object referring to the constant unrelated to the object motion

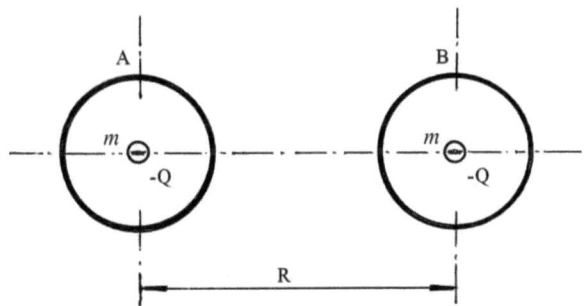

Fig. 4: Gravitational Mass Unrelated to the Object Motion

Object A and Object B have the same mass m and carry the equivalent negative charge -Q (as shown in Fig. 4), and the distance between Object A and B will be expressed as R. If the mutual universal gravitation between A and B is exactly equivalent to the mutual repulsion generated by the negative charge -Q contained in A and B, the universal gravitation and repulsion between A and B will be in equilibrium, and the distance R between A and B will be invariable and constant.

According to the Force Equilibrium Invariance Axiom, an observer being in motion at any velocity can observe that A and B will also inevitably be in equilibrium. It is unlike for the observer to observe that A and B will come into collision with each other or gradually separate from each other. It is only possible to meet this condition that the observer can also observe the universal gravitation between A and B equivalent to the electrostatic repulsion between A and B. Thus, we can inevitably draw the following conclusion that the gravitational mass is equivalent to the carried charge, i.e. the constant being independent of the motion.

4.3 General relativity applying to an object being in motion at the low velocity

The gravitational mass will be deemed as the constant unrelated to the motion, while the inertial mass will be related to the velocity of motion. This means that, the gravitational mass and the inertial mass will be deemed as the physical quantities being of the different properties; the gravitational mass is not possibly equivalent to the inertial mass. However, if the motion velocity of an object is less than the velocity of light, e.g. the motion velocities of the Sun's planets, man-made satellites and other planets are less than 1/6000 of the velocity of light, then there is a very small difference the inertial mass of an object and the rest mass of an object; it can be deemed that the inertial mass of an object is equivalent to the rest mass of an object, i.e. the inertial mass can be approximately regarded as the constant; under this condition, there is a fixed proportional relation between the inertial mass and the rest mass; through a transformation coefficient, it can be deemed that the gravitational mass is equivalent to the inertial mass; like this, through a transformation coefficient, it can be deemed that the length (height of water column) is equivalent to the pressure;

under this condition, the equivalence principle will be tenable; the inference made according to the general relativity will be in line with the reality; for example, the inference on the perihelion precession of the Mercury, GPS positioning and other items as made according to the general relativity will be consistent with the reality. However, if the motion velocity v of an object is relatively high, the inertial mass $m' = m\sqrt{1 - v^2/c^2}$ is significantly greater than the gravitational mass; the equivalence principle will not be tenable; the inference made according to the general relativity will be naturally divorced from the reality. For example, it is not correct to judge according to the general relativity that Black Hole exists when an object is in motion at the high velocity. Obviously, if the gravitational mass is supposed to be equivalent to the inertial mass and an object is in motion at the high velocity, the gravitational mass will become large as the inertial mass and the velocity of an object come close to the velocity of light and the gravitational mass will become very large, resulting in the Black Hole thereby. If the gravitational mass is constant and invariant, the Black Hole will never exist naturally.

It can also be judged from the aspect of the material structure that the Black Hole will not exist.

According to the Black Hole theory, for a black-hole sphere with a diameter of 120m, its mass is equivalent to the total mass of four suns, i.e. the total mass of 1.3 million earths and the Earth's volume is 10^{15} times greater than that of the black-hole sphere with a diameter of 120m; in other words, the density of black-hole matter is 10^{21} times greater than that of the Earth, which means the mass of a black-hole matter about the same size as a grain of rice can be 1000 times greater than the total mass of all persons on the Earth (taking the average mass of a person at 50kg). It is well-known that all the matters in the universe consist of several elements among the 118 kinds of elements known. So far, no other cosmic matters have been found to make an exception yet.

However, according to calculations based on the density of black-hole matter, the atomic size of black-hole matter is equivalent to only $1/10^{21}$ of the atomic size of matter on the Earth. This is clearly impossible. The reason is that, if we assume the Black Hole is also composed of atoms, after all atoms of the Black Hole have collapsed, all the electrons around such atoms will fall on protons and change them into neutrons, and these neutrons will be gathered together to become a large-sized neutron, then the volume of such a large-sized neutron will be one million times greater than the theoretical volume of the Black Hole. Obviously, there is no possibility of such atoms in reality. Thus, it can be concluded that it is impossible for the Black Hole to be composed of real atoms; that is to say it is impossible for the Black Hole to exist as a real matter.

The general relativity can be deemed as the theoretical basis of Black Hole. The Black Hole cannot exist, which indicates that the general relativity will not apply to the case of an object being in motion at the high velocity. Thus, we can draw a conclusion from the above analysis that the general relativity will apply only when the motion velocity of an object is less than the velocity of light.

Chapter 5　Conclusion

The above discussion shows that the special relativity has no transformation formula of the basic physical force or the method to judge the operating speed and operation record of any moving object in the actual time. This paper describes what the immutable axiom of the force balance is, and creates the transformation formula of the force accordingly. Based on the experimental results of J.C.Hafele and R.E.Keating, the problem on the operating speed of the time is analyzed, and it is inferred that the time delay has two different natures, i.e. relative time delay and absolute time delay, and both are real delay. The relative time delay refers to the observation result obtained by observing the time variation of the other moving reference system from one reference system; and the absolute time delay refers to the actual time delay of the object in a circular motion, and by virtue of the absolute time delay, we can infer the operating speed and operation record of any moving object in the actual time. For the general relativity, it is thought that the acting force of the acceleration is different from the nature of universal gravitation, because the acting force of the acceleration is energy transfer force, and the universal gravitation is the field force. Based on the axiom of the immutable force balance, it is concluded that the gravitational mass is constant and is irrelevant to the motion; therefore, the equivalence principle is false in principle. Only when the motion speed of the object is very small relative to the light speed can the equivalence principle be approximately true and the analysis of the general relativity be in line with reality; however, when the motion speed of the object is relatively big, the gravitational mass has obvious difference with the inertial mass, the equivalence principle is false, and the inference of the general relativity is not correct. Taking the black hole for example: the black hole would not be there if the gravitational mass is constant; considering the analysis of the structure of matter, it is inferred that the black hole cannot be the real material composed of atoms, thus indicating that the inference made by the general relativity about the high-speed motion of the object is not in line with reality.

References

1.Wu Shoubo, Basics of Theory of Relativity, Shaanxi Science & Technology Press(1987)

2.Yao Kexin Set up Invariable Axiom of Force Equilibrium and Solve Problems about Transformation of Force and Gravitational Mass. Applied Physics Research. V01.5.No.1 February 2013

3.Electrodynamics compiled by Tsinghua University electrodynamics teaching and research group;China Higher Education Press(1978)

4.Yao Kexin A New Explanation of Deflection Results of Charged Particles in High-velocity Motion in Magnetic Field·Correction of Lorentz Force

APPLIED PHYSICS RESEARCH   Vol.7,No.5   2015

5.Yao Kexin Explanation of Electromagnetics by Motion of Electric Field LAP LAMBERT Academic Publishing(2015)

6.Internet-baidu·com (2014) Black Hole

INTERNATIONAL JOURNAL OF RECENT SCIENTIFIC RESEAR CH(IJRSR)

Volumc:6   Jssuc:8

Supplement 1   A New Explanation of Deflection Results of Charged Particles in High-velocity Motion in Magnetic Field·Correction of Lorentz Force

Abstract: It is incorrect to only apply mass change or time change in explanation of the deflection result of charged particle in high velocity motion in magnetic field. A scientific and correct method is to change mass and time at the same time. However, it is impracticable to necessitate force formula in simultaneous change of mass and time. The paper makes correction of Lorentz Force formula based on analysis method for acting force in electric field, and launches into a new understanding of deflection result of charged particles in high velocity motion in magnetic field according to the corrected Lorentz Force formula.

Keywords: Charged particle; Magnetic field; Deflection; Lorentz Force; Mass change; Time change; Electric field force

1. Introduction

It is known to all in the field of physics that charged particle moving through magnetic field shall undergo deflection. The general explanation of the phenomenon is that when a charged particle with a charge of Q moves through magnetic field at the velocity of V along the direction of x, Q shall undergo Lorentz Force $F = BQV$ (B—magnetic induction) along the direction normal to V. F shall make the charged particle with a mass of M and a charge of Q generate an acceleration a=F/M along the direction normal to V (direction of Y) to make Q generate migration Y along the direction normal to V, namely deflection Y. Suppose the time for Q to move through magnetic field is t, then the expression for deflection Y is as follows:

$$Y = \frac{1}{2}at^2 = \frac{F}{2M}t^2 \tag{1}$$

Where (1) is derived under the condition that the direction of F being unchanged; although when particle moves through magnetic field, directions of velocity V and F (F is normal to V) are changed to a certain extent, since t is too short, the changes in F and V are also minor, accordingly, we can consider that there is no change in the directions of F and V and that the error of Y may be ignored.

Where (1) is practical on condition that velocity V be low, but if V is very high, there is obvious deviation between the calculated result of (1) and measured result, higher V shall bring about larger deviation, experiment shows that when V is very high, deflection distance shall be:

$$Y = \frac{Ft^2}{2M}\sqrt{1 - \frac{V^2}{C^2}} \qquad \text{(c—light velocity)} \tag{2}$$

At present, there are several explanations for the experimental result indicated by expression (2), it is observed that all these explanations are unscientific, the reasons for which are as follows:

The most general explanation is that when charged particle is in high velocity motion, according to special relativity, the mass M of the charged particle shall increase to $M' = M/\sqrt{1-V^2/C^2}$, replace M in expression (1) with M' to obtain the practical expression (2).

Another method is to carry out analysis based on change of particle momentum. Suppose the migration velocity of particle along direction of Y is u and particle mass is $M' = M/\sqrt{1-V^2/C^2}$ and particle momentum along direction of Y is $P = M'u = Mu/\sqrt{1-V^2/C^2}$ and the acting force on particle along direction of Y is

$$F = \frac{dP}{dt} = \frac{M}{\sqrt{1-V^2/C^2}}\frac{du}{dt}$$

$$\frac{du}{dt} = \frac{F\sqrt{1-V^2/C^2}}{M}$$

$$Y = \int_o^t dt \int_o^t \frac{du}{dt} \cdot dt = \frac{Ft^2}{2M}\sqrt{1-V^2/C^2}$$

It is observed that du/dt in the analysis method is constant acceleration a in substance, the deflection it determined is still $Y = at^2/2$, and that the key to the analysis method is $M' = M/\sqrt{1-V^2/C^2}$, there is no substantial difference between this method and the first method. Compared with the first method, it is obvious that this method is lack of its physical significance.

There is another method specified in Berkeley Physics, the method defines that

$\Delta\tau$ is particle clock time and that mass is rest mass M. ΔY is displacement of particle in Δt along direction of Y, momentum: $P = M\,\Delta Y/\Delta\tau$. It is observed that the time for particle to undergo displacement along direction of Y is $\Delta\tau$, according to special relativity, there is $\Delta\tau = \Delta t\sqrt{1-V^2/C^2}$, therefore there is $P = M\,\Delta Y/\Delta\tau = M\,\Delta Y/\Delta t\sqrt{1-V^2/C^2}$. Since the displacement velocity of particle along direction of Y is $V = \Delta Y/\Delta t$, accordingly, $P = MV/\sqrt{1-V^2/C^2}$, the momentum given in the result is identical with that obtained in the previous method. It is natural to come to a conclusion identical with that obtained by expression (2) according to the deduction steps given in the previous method.

It is observed that the first and second methods only apply mass change, although they do not mention that they only adopt mass change, in fact, time is unchanged, being a constant; the third method only applies time change, specifying that mass is unrelated to motion, being a constant. However, according to special relativity, the mass of particle in high velocity motion and time are changed simultaneously, accordingly, it is improper to suppose mass is unchanged and time is unchanged, because it is in violation of special relativity and practical result.

In accordance with special relativity, the correct method in analyzing the deflection of grain is to use both mass conversion $M' = M/\sqrt{1-V^2/C^2}$ and time conversion $t' = t\sqrt{1-V^2/C^2}$. Take M' and t' into formula (1), and then

$$Y = \frac{Ft'^2}{2M'} = \frac{Ft^2}{2M}\left(1-\frac{V^2}{C^2}\right)^{3/2} \tag{3}$$

It is obvious that formula (3) is not in line with the result of (2). Why? It is certain that formula (1) is the basic formula of mechanics, and this shall be affirmed; besides, it is without doubt that quality and time shall be conversed at the same time. Therefore, the only possibility is that formula (3) does not coincide with reality, and it must be Lorentz Force, F that fails to show the actual power.

Lorentz Force $F = BQV$ is only the experimental conclusion that charged particles of low runner are stressed in uniform magnetic field. It doesn't have necessary theoretical explanation. To analyze Lorentz Force is in line with reality or not, it is necessary to know the reason and change rules of Lorentz Force. Therefore, we have to illustrate the root of Lorentz Force from the theory and then it can analyze and solve the actual problems in formula (3).

2. The original analysis of Lorentz Force

It is shown in experiments that electric charge will be influenced by force in electric field, and the electric field can be considered as the only origin of charge force. Therefore, the electric field force can be considered as the basic point in analyzing Lorentz Force.

It can be known from the superposition principle of static electric field that the distance between two positron and negatron in space, the total electric field of positron and negatron must be equal to the vector sum when the positron and negatron exist alone. It can be deduced that any free electron in conductor (hereinafter called negatron) and the proton that is equal with free electron in electric quantity (hereinafter called positron) has their own electric field. Therefore, when there is current in conductor, namely the macroscopic motion of negatron in conductor exists, the electric field of negatron must move with negatron. And the macroeconomic effect is that the negative electric field moving around conductor exists along with the static positive electric field. In terms of magnet, the electron that spins in same direction in magnetic domain is similar to the negatron of macroscopic motion in conductor. Therefore, there are negative electric field of macroscopic motion and static positive electric field in magnetic domain.

First of all, we shall analyze the relationship between magnetic field and electric field. It is showed in Biot-Savart theorem that, the feeling strength of any current element Idl (small lines on the wire that in the same direction of current) at the place of r is

$$dB = \frac{\mu_0}{4\pi} \cdot \frac{Idl}{r^2} \times r_0 \qquad \text{(Unit vector in the direction of } r_0 \text{ - } r \text{)}$$

Since I is actually formed by free electron in the wire in the speed of V (in the opposite direction of current I). Taking $-\tau$ as the linear charge density in the wire, then $\quad Idl = -\tau(-V)dl = dQV \qquad$ (the quantity of free electron in dQ - dl) and because $\mu_0\varepsilon_0 = 1/C^2$, $\quad K = 1/4\pi\varepsilon_0$, then

$$\frac{\mu_0}{4\pi} = \frac{K}{C^2}$$

$$dB = \frac{K}{C^2} \cdot \frac{dQV}{r^2} \times r_0 = \frac{1}{C^2} V \times K \frac{dQ}{r^2} r_0$$

$$= \frac{1}{C^2} V \times dE = \mu_0\varepsilon_0 V \times dE \ .$$

In the upper formula, dE is the electric field intensity of dQ at the place of R. and the formula can be written as

$$B = \frac{1}{C^2} V \times E = \mu_0 \varepsilon_0 V \times E \qquad (4)$$

(4) The formula shows that the feeling strength at some place is equal to the vector product of kinematic velocity V and the electric field intensity. The formula shows that magnetic field is a moving electric field.

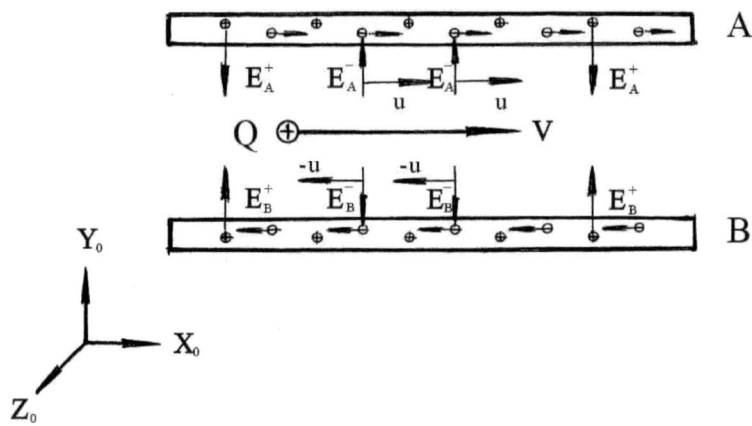

Figure 1 Origin analysis of Lorentz Force

It is defined that flat consisting of wires with current is the current surface, in Figure 1, A and B indicates two sections of "infinity" current surface. A and B current surfaces are made up of two common wire with current, and the current of A and B is equal but in opposite direction. In Figure, E_A^- and E_B^- shows the electric field intensity that negatron electron in motion in current surface; E_A^+ and E_B^+ indicate the electric field intensity that produced by static positive electron that is equal to negatron in electric quantity.

It shall count out the feeling strength between two current surfaces A and B. It can be found out from formula (4) that, B is originated from moving electric field. The motion of E_A^+ and E_B^+ has nothing to do with B. And then we only analyze the electric field of E_A^- that moves in the speed of V and then electric field of E_B^- in the speed of –V. Obviously, E_A^- and E_B^- is equal but in opposite direction, namely $E_A^- = E_B^-$. where E_0 is the absolute value of E_A^- and E_B^-.

It shall count out the feeling strength of B_A from E_A^-, and from formula (4)

$$B_A = \mu\varepsilon_0 u \times E_A^- = \mu_0\varepsilon_0 u E_0 Z_0 \quad (Z_0 \text{ is the unit vector in the direction of Z})$$

The feeling strength B_B produced by E_B^- is

$$B_B = \mu_0\varepsilon_0 - u \times E_B^- = \mu_0\varepsilon_0 - u \times \left(-E_A^-\right) = \mu_0\varepsilon_0 u E_0 Z_0$$

The total feeling strength produced by current surface A and B is

$$B = B_A + B_B = 2\mu_0\varepsilon_0 u E_0 Z_0$$

Figure 1, Q is the charge particle with positive electricity in the speed of V paralleling two electricity surfaces. The Lorentz Force that is bore by grant with Q charge is

$$F = QV \times B = 2\mu_0\varepsilon_0 Vu E_0 Q(-Y_0) \quad (Y_0 \text{ is the unit vector in the direction of Y)} (5)$$

When we make analysis of the source of Lorentz Force, it will be our natural selection for us to start with the source of force Q. There is no doubt that the only source of Q force is under the effect from electric field. There are not only electric field E_A^- and E_B^- but E_A^-、 E_B^+ as well between the two current surface A and B; because the current of E_A^+ and E_B^+ is the same intensive with contrary direction and they has the same relative velocity to any moving Q, E_A^+ and E_B^+ will have a constant force on force Q with contrary directions and their forces will be cancelled out, namely the effect of E_A^+ and E_B^+ can not be considered on force Q. Without consideration of E_A^+ and E_B^+, the force of Q is only sourced from E_A^- and E_B^- possibly.

Let us study the force on Q from E_A^- firstly, it is can be known that the relative velocity of E_A^- and Q is $V_A = V - u_0$ from Figure 1; According to Special Relativity, the electric field E_A^- with V_A moving on will shrink in the proportion of $1/\sqrt{1 - V_A^2/C^2}$ in the direction of V_A and the field density is inversely proportional to the distance of power line, therefore the filed density shall be magnified as:

$$E_A^{-'} = E_A^- / \sqrt{1 - V_A^2/C^2} = E_A^- / \sqrt{1 - (V - u)^2/C^2}$$

According to binomial series, $1/\sqrt{1 - X^2}$ can be expanded into $1 + X^2/2 + 3X^4/8 + 5X^6/16 + \ldots$, when x is relative small, it can be neglected including X^4 and $X^6 \ldots$ etc., it is $1/\sqrt{1 - X^2} = 1 + X^2/2$. Under general condition, V and u is far less than C, therefore $E_A^{-'}$ in the above formula can be converted into:

$$E_A^{-'} = E_A^- \left[1 + (V - u)^2/2C^2\right] = E_0 \left[1 + (V - u)^2/2C^2\right] Y_0$$

Let us turn to E_B^-, it is can be known that the relative velocity of E_B^- and Q is $V_B = V + u$ and it can be got from the explanations mentioned above as:

$$E_B^{-'} = E_B^- / \sqrt{1 - (V + u)^2/C^2} = E_0 \left[1 + (V + u)^2/2C^2\right](-Y_0)$$

For the moving Q, the total field intensity between the current surface A and B is

$$E' = E_A^{-'} + E_B^{-'}$$

$$= E_0 \left[1 + \frac{(V - u)^2}{2C^2}\right] Y_0 + E_0 \left[1 + \frac{(V + u)^2}{2C^2}\right](-Y_0)$$

$$= E_0 \left[1 + \frac{(V+u)^2}{2C^2} - 1 - \frac{(V-u)^2}{2C^2} \right] (-Y_0)$$

$$= \frac{2}{C^2} VuE_0(-Y_0) = 2\mu_0\varepsilon_0 VuE_0(-Y_0)$$

The force on Q is:

$$F = E'Q = 2\mu_0\varepsilon_0 VuE_0 Q(-Y_0)$$

It is obvious that the formula is just the Lorentz Force Formula(5) mentioned before. It can be concluded from the above analysis that: when Q is moving in magnetic field at the velocity V, one moving electric field with two current surface E_A^- and E_B^- will be formed; for Q, their velocity are $V+u$ and $V-u$ *respectively and the electric field of V+u will shrink greater with more intensive field*; when the electric filed of $V-u$ shrinks with less intensive filed intensity, one synthesis electric field will be formed by their differentiated filed intensity, whose force on Q will be just Lorentz Force.

3. Correction of Lorentz Force

It can be known from the analysis on the derivation formula of above-mentioned Lorentz Force that the derivation process will not hold when the velocity of Q is very large (almost reaching the velocity of light C); there are two reasons, one is that when V is very large, V^4/C^4 and V^6/C^6 ...... can not be neglected, and the above derivation will not hold naturally; the other is that when V is very large, its velocities can not be added or subtracted directly, namely $V_A \neq V-u$ and $V_B \neq V+u$; therefore, the above derivation will not hold.

Now, let us make further analysis on the force of charged particle in high velocity motion (expressed with electric charge Q) in accordance with the synthesis velocity method given in Special Relativity.

The synthesis velocity mentioned in Special Relativity refers to that in the direction of X, Y and Z. For the condition given in Figure 1, the velocity on direction Z does not exist and migration velocity on direction Y is very slow, that's to say they can be neglected owing to the comparison with V. Therefore, it can be considered that there is also no velocity component on direction Y and we only need to make analysis of the synthesis velocity on direction X whose formula is given as:

$$V_x = \frac{u_x + v}{1 + u_x v/C^2} \tag{6}$$

In the formula, v is the moving velocity of the other inertial system Z' observed from inertial system Z with the motionless of relative subject M; u_x is the moving velocity relative to subject N, V_x is the relative velocity speed of M and N. For Figure 1, we would like to conclude the force on Q (subject M), namely Q is of inertial system of Z and two current surface is of the inertial system of Z', The velocity of Z' (two current surface) observed from Z' is $-V$, namely $v = -V$, and the motion velocity of E_A^- (subject N1) observed from Z' (two current surface) is u, namely $u_x = u$; The velocity speed of Q (subject M) relative to E_A^- (subject N1) is in accordance with formula (6)

$$V_A = \frac{u - v}{1 - uv/C^2} \tag{7}$$

However, the motion velocity of E_B^- observed from Z' (two current surface) is $-u$, namely $u_x = -u$; The velocity speed of Q (subject M) relative to E_B^- (subject N1) is in accordance with formula (6)

$$V_B = \frac{-u-v}{1+uv/C^2} \qquad (8)$$

E_A^- moving at the speed of V_A will shrink for Q, it shall be magnified as:

$$E_A^{-'} = E_A^- / \sqrt{1-V_A^2/C^2}$$

Put V_A concluded from (7)in the following formula:

$$E_A^{-'} = \frac{E_A^-}{\sqrt{1-\left(\frac{u-V}{1-uV/C^2}\right)^2 / C^2}} = \frac{E_A^-}{\sqrt{1-\frac{C^4(u-V)^2}{(C^2-uV)^2} / C^2}}$$

$$= \frac{(C^2-uV)E_A^-}{\sqrt{(C^2-uV)^2-C^2(u-V)^2}} = \frac{(C^2-uV)E_A^-}{\sqrt{C^4+u^2V^2-C^2u^2-C^2V^2}}$$

$$= \frac{(C^2-uV)E_A^-}{\sqrt{(C^2-V^2)(C^2-u^2)}} = \frac{(C^2-uV)E_A^-}{C^2\sqrt{(1-V^2/C^2)(1-u^2/C^2)}}$$

The movement velocity of negative electron is far smaller than C (electron in magnetic domain is also smaller) in the moving electric field formed in two current surfaces A and B; therefore u^2/C^2 is very small, which can be reckoned as $1-u^2/C^2 \approx 1$, and then the above formula can be converted into:

$$E_A^{-'} = \frac{(C^2-uV)E_A^-}{C^2\sqrt{1-V^2/C^2}} \qquad (9)$$

For E_B^-, E_B^- moving at the speed of V_B shall be shrank and magnified as:

$$E_B^{-'} = E_B^- / \sqrt{1-V_B^2/C^2}$$

Put the V_B got in formula(8)into the following formula:

$$E_B^{'} = \frac{E_B^{-}}{\sqrt{1-\left(\dfrac{u+V}{1+uV/C^2}\right)^2 \Big/ C^2}} = \frac{E_B^{-}}{\sqrt{1-\dfrac{C^4(u+V)^2}{(C^2+uV)^2}\Big/ C^2}}$$

$$= \frac{(C^2+uV)E_B^{-}}{\sqrt{(C^2+uV)^2 - C^2(u+V)^2}} = \frac{(C^2+uV)E_B^{-}}{\sqrt{C^4+u^2V^2-C^2u^2-C^2V^2}}$$

$$= \frac{(C^2+uV)E_B^{-}}{\sqrt{(C^2-V)(C^2-u^2)}} = \frac{(C^2+uV)E_B^{-}}{C^2\sqrt{1-V^2/C^2}} \qquad (10)$$

The field intensity felt by Q is:

$$E^{'} = E_B^{'} + E_A^{'}$$

$$= \frac{(C^2+uV)E_B^{-}}{C^2\sqrt{1-V^2/C^2}} + \frac{(C^2-uV)E_A^{-}}{C^2\sqrt{1-V^2/C^2}}$$

Owing to $E_A^{-} = -E_B^{-}$, the above formula can be converted into:

$$E^{'} = \frac{(C^2+uV)E_B^{-}}{C^2\sqrt{1-V^2/C^2}} - \frac{(C^2-uV)E_B^{-}}{C^2\sqrt{1-V^2/C^2}}$$

$$= \frac{2uVE_B^{-}}{C^2\sqrt{1-V^2/C^2}} = \frac{2\mu_0\varepsilon_0 uVE_0}{\sqrt{1-V^2/C^2}} \cdot (-Y_0) \qquad (11)(\text{ The first capitalized E, bold})$$

For the sake of telling the force F, namely Lorentz Force F, exerted by electric field when Q is moving at medium and low speed, the force F_V on Q exerted by electric field can be concluded from formula (11):

$$F_V = E'Q = \frac{2\mu_0\varepsilon_0 VuE_0 Q}{\sqrt{1-V^2/C^2}}(-Y_0) \qquad (12)$$

It is shown from former formula (5) that Lorentz Force on Q is $F = 2\mu_0\varepsilon_0 VuE_0 Q(-Y_0)$ when it is moving at medium and low speed, and put it into formula (12):

$$F_V = \frac{F}{\sqrt{1-V^2/C^2}} = \frac{QV \times B}{\sqrt{1-V^2/C^2}} \tag{13}$$

The formula (13) can be used for the correction of the formula of Lorentz Force. It is obvious that formula (13) is also qualified for that at both high and low speed when Q is moving in magnetic field at medium and low speed.

4. Interpretation of charged particles' deflection with the help of corrected formula for Lorentz Force

The corrected Lorentz Force, F_V, on Q got from above analysis is just the electric field force, which is relatively static to Q; it will move together with Q at the same speed V. For the observer with static relative magnetic field (two current surfaces), force F_V is a moving one. I have mentioned in "Set up Invariable Axiom of Force Equilibrium and Solve Problems about Transformation of Force and Gravitational Mass" (References. 3) that the force on moving subject will also change like its length, mass and time. The conversion formula for the force moving at the speed V is:

$$F' = F\sqrt{\frac{1 - cos^2\theta V^2/C^2}{1-V^2/C^2}} \tag{14}$$

In the formula, θ is the included angle between F and V.

For F_V analyzed in Figure 1, the included angle between F_V and V is $\theta = 90°$ and $cos\theta = 0$, therefore the force F_V' can be converted into in accordance with the formula(14):

$$F_V' = F_V / \sqrt{1-V^2/C^2}$$

Put formula (13) into this formula (13)

$$F_V' = \frac{F}{1 - V^2/C^2}$$

F_V' is just the real force exerted on charged particle we would like to discuss, put real F_V' into formula (3) instead of F:

$$Y = \frac{F_V' t^2}{2M}\left(1 - \frac{V^2}{C^2}\right)^{3/2} = \frac{Ft^2}{2M}\sqrt{1 - V^2/C^2}$$

It is obvious that the deflection distance Y is just as determined in experiment room, namely true deflection distance formula (2).

It is shown from the above analysis that the problem in connection with the deflection of charged particle moving at high speed can be solved reasonably provided that its mass and time are changed at the same time with the help of corrected Lorentz Force in combination with force transformation.

5. Making analysis of particles' deflection from the inertial system of relative static of charged particles

According to the relativity principle given in Special Relativity, for the observers of inertial system Z (inertial system Z' moving at the same speed of Q) and inertial system Z' (relative magnetic field or static surface), it is of no doubt that they will have the same result of the deflection on Q passing through the magnetic field on direction Y.

The former analysis is the conclusion of the observer relatively static to Z' , for this conclusion, the real force F_V' is got with the help of corrected Lorentz Force and changed combined power, moreover the analysis result compliance with fact is concluded. Z is moving relatively to Z' at the speed V, it can be reckoned that F_V is static relatively to Z (with slight movement of F_V on direction Y neglected) there is no need for changing F_V any longe; Whether the change of this kind of conditions will have any influence on our analysis of the corresponding deflection? Let us make detailed analysis for it.

For observer Z, Q is static and one magnetic field (inertial system Z') is passing through Q at the speed $-V$, one downward force F_V is be exerted on Q, and moreover distance Y is moved downwards (deflection); because the speed of Q on direction Y is very slow, the mass of Q is still static one, namely no change is occurred to it. Besides the force on Q is also a static one, namely $F_V = F/\sqrt{1 - V^2/C^2}$ (corrected Lorentz Force), of course F_V will not change. The accelerated speed of Q on direct Y is $a = F_V/m$. Provided the time Q passing through the magnetic field is t_V observed by Z, the deflection distance of Q on direction Y is:

$$Y = \frac{1}{2}at_V^2 = \frac{Ft_V^2}{2m\sqrt{1 - V^2/C^2}} \qquad (15)$$

What is the relationship between t_V and Z' observed by Z? First of all, we shall make sure that the absolute moving speed V and $-V$ of Z and Z' is the same and Z observes t is the time Q passing the magnetic field at the speed V, namely $t = L/V$; for Z, the magnetic in L wide is moving at the speed of $-V$, according to the Relativity Principle given in Special Theory, the width of the magnetic field moving at the speed $-V$ is $L' = L\sqrt{1 - V^2/C^2}$; therefore the time used by the magnetic passing through Q from the observation of Z is $t_V = L'/V = L\sqrt{1 - V^2/C^2}/V = t\sqrt{1 - V^2/C^2}$, put it into formula(15),

$$Y = \frac{Ft^2}{2m\sqrt{1-V^2/C^2}}\left(1-V^2/C^2\right) = \frac{Ft^2}{2m}\sqrt{1-V^2/C^2}$$

It is obvious that the relationship formula is the same as real deflection distance (2).

For the above analysis, the deflection results of charged particles moving at high speed got by Z' and Z shall be the same, which is in compliance with Relativity Principle.

6. Application scope of Lorentz Force

I hereby point out that the force on Q will be the same only in the uniform magnetic field through the analysis with the help of electric magnetic field and Lorentz Force; under normal condition, the two results derived from the two methods will not be the same in non-uniform magnetic field.

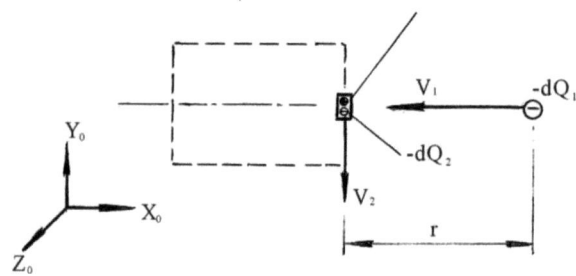

Figure 2 Analysis on interaction between moving electric charge and current component

Figure 2, electric charge $-dQ_1$ is moving towards Idl at speed V_1, the capacity of negative electron moving in Idl is $-dQ_2$ and velocity speed of $-dQ_2$ is V_2 (contrary direction with I), the capacity of positive electron in Idl with the same capacity of that in $-dQ_2$ is $-dQ_2$.

First, the interaction force between $-dQ_1$ and Idl is subject to analysis based on electric field. Since the relative velocity between negative charges or negative charge and positive charge in any relative motion is identical in magnitude and contrary in direction, The shrinkage condition observed each other is completely similar; therefore the force on each other is the same as that when they are relatively static. It is certain that they are the same in force and contrary in direction. We can know from this that both the interaction between $-dQ_1$ and $-dQ_2$ and that between $-dQ_1$ and positive electron dQ_2 are equal in force and contrary in direction; namely, the inter-force is naturally equal with contrary direction. It is obvious that the analysis result for electric field is in compliance with reaction law.

It is observed from analysis of interaction between $-dQ_1$ and IdL based on Lorentz Force that only $-dQ_2$ in motion in current element generates magnetic field; since dQ_2 is in no motion, accordingly, it generates no magnetic field.

The electric field strength generated by $-dQ_2$ in $-dQ_1$ is dE_2, the following X_0, Y_0 and Z_0 are unit vectors, the motion velocity of dE_2 is V_2, V_2 is referred to as $V_2(-Y_0)$, according to the previous formula(4), the magnetic induction strength generated by $dE_2(-X_0)$ in $-dQ_1$ is as follows:

$$dB_2 = \frac{1}{C^2}V_2 \times dE_2 = \frac{1}{C^2}V_2(-Y_0) \times dE_2(-X_0)$$

$$= \frac{1}{C^2} V_2 dE_2(-Z_0)$$

The motion velocity of $-dQ_1$ is V_1, V_1 is also referred to as $V_1(-X_0)$, accordingly, Lorentz Force on $-dQ_1$ is as follows:

$$dF_1 = -dQ_1 V_1 \times dB_2 = \frac{-dQ_1}{C^2} V_1(-X_0) \times V_2 dE_2(-Z_0)$$

$$= \frac{-dQ_1}{C^2} V_1 V_2 dE_2(-Y_0) = \frac{1}{C^2} V_1 V_2 dE_2 dQ(Y_0)$$

With regard to Lorentz Force on $-dQ_2$, the electric field strength generated by $-dQ_1$ in $-dQ_2$ is $dE_1(X_0)$, the magnetic induction strength generated by $dE_1(X_0)$ in $-dQ_2$ is as follows:

$$dB_1 = \frac{1}{C^2} V_1(-X_0) \times dE_1(X_0) = 0$$

Since positive charge dQ_2 in no motion, therefore dQ_2 is not subject to Lorentz Force, the Lorentz Force on current element IdL is the Lorentz Force on $-dQ_2$, the Lorentz Force on $-dQ_2$ is as follows:

$$dF_2 = -dQ_2 V_2(-Y_0) \times dB_1 = 0 \qquad (dB_1 = 0)$$

It is observed from comparison between action dF_1 and reaction dF_2 that they are different in magnitude and identical in direction; that is to say that adopting Lorentz Force to analyze interaction force between charge in motion and current element results in action and reaction are different in magnitude and not contrary in direction; namely, the result is contrary to law of reaction. It is observed from further analysis that in addition to V_1 in parallel with V_2 as motion velocity of V_1, it is observed from analysis of Lorentz Force along any direction of V_1 that the interaction and reaction between $-dQ_1$ and Idl are different in magnitude and not contrary in direction, namely the results are contrary to the law of reaction.

It is stated that there is no isolated current element in the universe and it is impractical to adopt current element for analysis. It is also known that scientific theory necessitates both tenable macro-analysis and micro-analysis. In fact, many theories are based on from micro-analysis to macro-analysis. If micro-analysis fails to be tenable, there shall be an example showing that macro-analysis fails to be tenable. The micro-current element given in Figure 2 extends along the dotted line shown in the fig, becoming the macro-current carrying coil. Analysis of the interaction force between current carrying coil and $-dQ_1$ shown in dotted line in fig.2, the inevitable analysis conclusion is that the interaction and reaction between charge in motion $-dQ_1$ and current carrying coil are not contrary in direction and different in magnitude. Reference document 2 details the analysis result.

According to the above mentioned analysis, it is concluded that Lorentz Force is only applicable to uniform magnetic field and not applicable to non-uniform magnetic field. In order to analyze the force on charge in motion in non-uniform magnetic field, the only way is to adopt acting force in electric field for analysis.

References

1. Yao Kexin Inferring the Fact that Static Magnetic Field Exists Along with Electrostatic Field and Conducting Experimental Verification in Accordance with the Theory of Relativity. Applied Physics Research Vol.4.No.1. February 2012.

2. Yao Kexin Question on Some Principles of Electromagnetism. Applied Physics Research V01.4.No.3August 2012

3. Yao Kexin Set up Invariable Axiom of Force Equilibrium and Solve Problems about Transformation of Force and Gravitational Mass. Applied Physics Research. V01.5.No.1 February 2013

4. Yao Kexin Explanation of Electromagnetics by Motion of Electric Field LAP LAMBERT Academic Publishing(2013)

5. E.M. Purcell [U.S.] Berkeley Physics Course, published by Science and technology of China press (1979)

6. Fundamental of Relativity Theory, compiled by Wu Shoubo, published by Shaanxi Science & Technology Press (1987)

APPLIED PHYSICS RESEARCH   Vol.7,No.5   2015

Supplement 2 Actual Time Dilation is Dependent on Absolute Velocity

It is Imp

Ossible for Human Being to have Time Travel

Abstract: A description of relative time dilation and absolute time dilation are given and it is pointed out that confusion of the two terms is likely to lead to paradox and it is considered that circular motion velocity to center of a circle is absolute velocity and the time dilation due to absolute velocity is actual dilation---absolute dilation. Absolute dilation is unrelated to any object in relative motion. The time dilation in correlated circular motion has been discussed to conclude that the absolute velocity human being can achieve is very limited; therefore, it is impossible to achieve time travel.

Key words: Special relativity theory, relative velocity; relative dilation, absolute speed; absolute dilation, circular motion, correlated circular motion, time travel

1. Introduction

Special relativity theory points out that clock time in motion is to dilate (lose time), accordingly, for two clocks A and B in relative motion, A considers that clock time B in motion dilated and B also considers that clock time A dilated. We define the time dilation mutually considered by A and B to be the other time dilation as relative dilation. It is obvious that we can not determine which one actually dilated between A and B based on relative dilation and that we can not explain some question about time dilation in a scientific way if we take relative dilation as actual dilation, for example: question about twin paradox.

At present, all the explanations of twin paradox are illogical. One explanation is that although twins A and B are identical in terms of relative motion speed, A in space travel shall experience an accelerating process to get away from earth, this accelerating process shall make A younger than B, another explanation is that A in space travel shall experience a rotating process to return to earth, this rotating process shall make A younger than B (See reference document 3 for above mentioned content) . It is observed from analysis that these two explanations are illogical, the reason for which is that A in space travel for one day and for ten years shall be identical in terms of accelerating process and rotating process, it is observed from above mentioned two explanations that A in space travel for one day and for ten years shall be identical in terms of young effect in substance. It is obvious that the inference is false, the cause for which is that there is no clear difference between relative time dilation and actual time dilation, the paper presents two concepts namely absolute velocity and absolute dilation to solve the problem.

2. Actual time dilation is unrelated to relative velocity

First, we shall verify that relative time dilation is unrelated to actual time dilation in theory, as shown in the following fig:

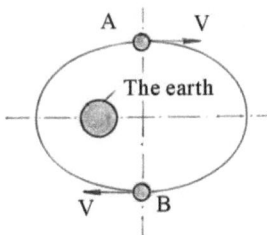

Fig.: Spacecrafts A and B are in relative motion

We launch a spacecraft A from the ground to fly around an elliptical orbit, the time for A to make a complete revolution around the earth is indicated by T, after T/2, we launch another spacecraft B from the same location to fly around the same orbit, having the same orbital period. It is obvious that A (Or B) finds B (Or A) is in motion relative to it at any instant. According to special relativity theory, both A and B speculate that the other clock lost time (Relative dilation). After a period of time, B and A return to earth at intervals of T/2. Since A and B have identical flight process, if there is time dilation in their clocks, it is certain that their time dilation shall be identical, therefore, after return to earth, it is impossible for A and B to find that the other clock lost time, that is to say that the speculation by A and B indicating the other clock lost time (Relative dilation) is not consistent with the fact, which is to say that it is impossible to know the actual time dilation only according to relative dilation.

It is also observed in this way that for the objects different in motion velocity such as M_1, M_2, M_3..., their motion velocity of M under observation relative to them is V_1, V_2, V_3... respectively, according to special relativity theory, M_1, M_2, M_3、 ..., it is speculated that the time dilation of M is t_1, t_2, t_3.... Since it is impossible for clock dilation of M is equal to t_1, t_2, t_3、 ...at the same time, thereby, It is impossible for time dilation of M clock is associated with object in motion relative to it.

It should be pointed out that shortened length and time dilation of an object in motion concluded by special relativity theory are the results of transformation of physical parameters in two different reference systems and the difference between different reference systems in measurement and observation, which do not mean that the length of the object in motion is shortened and that the clock lost time. For example, we find a person standing on a hill becomes smaller in stature, it does not mean that the person really becomes smaller in stature.

3. Absolute velocity and absolute dilation

One man holds a rope connected to a heavy M in his hand, he tries to make the rope connected to M turn round his hand, any one in the car or on the plane can see that M is rotating round his hand and never think that his hand is rotating relative to M and it may be essential that M itself also thinks that it is rotating round hand, just like the earth rotates round the sun, believed by everyone on earth. Accordingly, the motion velocity of M relative to hand (Including all stationary objects) is absolute not relative. The circular motion velocity of M relative to hand is defined as absolute speed.

If the time for M to turn from beginning to end under the observation by people on the ground is indicated by T and the rotating velocity of M is indicated by V, according to special relativity theory, the time interval T' between clock on M and T

$$T' = T\sqrt{1 - V^2/C^2}$$ (C—light velocity)   (1)

T' is the result of observation of stationary hand and M in motion, being generally accepted, which is different from relative dilation at root, relative dilation means every one considers that the other clock lost time and that the clock of his or her own did not lose time, being not accepted, thereby, the generally accepted time dilation T' in circular motion is defined as absolute dilation.

Similar to the rotating rope, the earth rotates round earth axis (North and south poles), the rotating velocity of every point on earth to earth axis is its absolute speed. It is obvious that the absolute velocity on equator is the highest and the absolute dilation thereon is also the largest, the absolute velocity on north and south poles is zero. (The earth axis rotates round the sun, in which there is absolute velocity of the earth axis to the sun, as shown in note 5), the clock on the north and south poles is the standard clock to indicate motion velocity in the time frame on earth.

4. Experiment by J·C·Hafele and R·E·Keating

In 1971, J·C·Hafele and R·E·Keating made an experiment on relation between time dilation and motion velocity. They installed four caesium atomic clocks on the plane over equator, when the plane revolved round the earth from east to west along equator, they found that the average reading of the four caesium atomic clocks gained time by 273×10-9 (Showing negative dilation) compared with that of the caesium atomic clocks on the ground; and when the plane revolved round the earth from west to east along equator they found that the average reading of the four caesium atomic clocks lost time by 59×10-9 (Reference document 1) compared with that of the caesium atomic clocks on the ground, what is the reason for the results? Since the earth revolves from west to east, when the plane is flying west, the absolute velocity of the plane is the difference between absolute velocity of the ground along equator and the flight velocity of the plane. The absolute velocity of the plane and the absolute dilation of the clock thereof are smaller than the absolute velocity of the ground along equator and the absolute dilation of the clock thereof, therefore, the caesium atomic clocks on the plane gain time compared with the caesium atomic clocks on the ground.

According to relativity theory, they realize by calculation that the caesium atomic clocks on the plane gains time by 275×10^{-9} seconds compared with those on the ground, which is in accordance with the actual reading by and large; when the plane is flying east, the absolute velocity of the plane is the sum of the absolute velocity of the equator and the flight velocity of the plane, the absolute velocity and absolute dilation of the plane are greater than the absolute velocity and absolute dilation of the equator. Accordingly, the caesium atomic clocks on the plane lose time compared with those on the ground. Their theoretically calculated value indicates that the caesium atomic clocks on the plane lose time by 40×10^{-9} seconds compared with those on the ground. It is obvious that there is marked deviation between the theoretical value and actual 59×10^{-9} seconds in word, but it should be noted that these two values are not absolute dilation.

In the experiment, the absolute dilation of the caesium atomic clock on the plane is the sum of the absolute dilation of the caesium atomic clock of the equator and 59×10^{-9} seconds; and that the theoretically calculated value shall be the sum of absolute dilation of the caesium atomic clock and 40×10^{-9} seconds. Since there is no caesium atomic clock for north pole or south pole in the experiment, not knowing the absolute dilation of the caesium atomic clock of the equator, according to conservative estimate, its value shall be more than 500×10^{-9} seconds, based on which, it is estimated that the numerical error between the above mentioned absolute dilation $(500 \times 10^{-9} + 40 \times 10^{-9})$ and $(500 \times 10^{-9} + 59 \times 10^{-9})$ shall be within 3.4%, accordingly, it is considered that the theory is consistent with reality. According to the above mentioned experiment results, the results of the experiment by J·C·Hafele and R·E·Keating are acceptable, verifying that the time dilation theory of clock in motion given in special relativity theory conforms to objective reality and also proving that absolute dilation is dependent on absolute velocity to the full extent and unrelated to relative velocity of any object.

And now, we can explain twin paradox in a simple way, the result is whether twin A in space travel is younger than B is dependent on whether the absolute dilation of A is larger than the absolute dilation of B to the full extent, take the above mentioned experiment as an example, if A flies from east to west by plane, then A is not younger than B; if A flies from west to east by plane, then A is younger than B.

5. Absolute dilation of correlated circular motion

The earth revolves round the sun, and the moon revolves round the earth, if the moon has artificial satellite, the satellite revolves round the moon, the correlated rotation motion is defined as correlated circular motion.

The absolute velocity for the earth to revolve round the sun is indicated by V, its time interval is indicated by T; the absolute velocity for the moon to revolve round the earth is V_m, the time interval relative to V_m and T is indicated by T_m; the absolute velocity for the moon   satellite to revolve round the moon is V_s, the time interval relative to V_s and T is indicated by T_s. Since we carry out observation and analysis based on the human oriented point of view, thereby we define T (time interval of earth axis)as reference time interval. According to special relativity theory, the formulae are as follows:

$$T_m = T\sqrt{1-V_m^{\,2}/C^2} \tag{2}$$

$$T_S = T_m\sqrt{1-V_S^2/C^2} = T\sqrt{1-V_m^2/C^2}\cdot\sqrt{1-V_S^2/C^2} \tag{3}$$

If the time interval of T relative to the sun is indicated by T_0, then there is

$$T_0 = T\Big/\sqrt{1-V^2/C^2} \tag{4}$$

If the velocity for a planet of the sun to revolve round the sun is V_x, the time interval relative to V_x and T is indicated by T_x, then there is

$$T_x = T_0\sqrt{1-V_x^{\,2}/C^2} = T\sqrt{1-V_x^2/C^2}\Big/\sqrt{1-V^2/C^2} \tag{5}$$

It should be pointed out that the orbits for earth and moon and other planets of the sun are not pure circular orbits, since there is a radius of curvature at every point on the ellipse, namely every one has a corresponding center of a circle, consequently, the motion velocity of the planet at every point on the elliptical orbit is absolute velocity.

6. It is impossible for humane being to have time travel

Synchrocyclotron is able to accelerate the motion velocity of particle to the velocity approximate to light velocity; can the motion velocity of the manned spacecraft be accelerated to the velocity approximate to the light velocity by the corresponding synchrocyclotron? The answer is no, the reason for which is that it is difficult for human being to fabricate practical and grand synchrotron for spacecraft and the more important is that human being is not charged body and that it is impossible for human in synchrotron to endure huge centrifugal force due to circular motion because human being have no Lorentz force to offset centrifugal force. In particular, human can not withstand the centrifugal force as a result of the $V = 30km/s$, being ten thousandth of light velocity, let along light velocity. It is observed from calculation that if the mass of a man $M = 50kg$ and the radius of cyclotron $R = 50km$, the centrifugal force (mv^2/R) to be withstood by the man is up to 900000 N, being 1800 times the weight of the man (About 500 N), in this way, the time dilation achieved by the man within a year is only 0.16 second, therefore, there is no time dilation, namely it is impossible for human to achieve time travel by cyclotron.

Since the universal gravitation among celestial bodies (Mass is indicated by M) shall offset the centrifugal force due to circular motion, thus, it is practical for human to achieve absolute dilation by revolving round the celestial body in spacecraft. Assume the mass of a man is m and the revolving radius of M is R, the velocity is V, when revolving of m round M is in stable equilibrium state, the centrifugal force mv^2/R shall be equal to universal gravitation GMm/R^2 (G —universal gravitation constant), based on which the formula is as follows:

$$V = \sqrt{GM/R} \tag{6}$$

It is observed from formula (6) that the greater M/R is, the higher V will be. On the premise that M remains unchanged, the smaller R is, the higher V will be.

The known flight velocity for spacecraft to revolve round the earth is up to 7.9m/s, it is observed from formula (1) that its absolute dilation is 0.01 second per year. If spacecraft revolves round the moon, it is observed from putting the mass and radius of the moon into the formula (6) that the maximum velocity of the spacecraft is no more than1.7m/s and the velocity for the moon to revolve the earth is 1.02m/s, it is observed from formula (3) that the absolute dilation T_s of the spacecraft is 0.0052 second within one year; the velocity for Mercury to revolve the sun is the highest among planets of the sun, but its temperature is too high; Venus is in the second place, assume that man is to revolve round the sun along Venus orbit by spacecraft at the velocity of 34km/s. the average velocity for the earth to revolve round the sun is 29.78km/s, put the two velocities into formula(5)to calculate the spacecraft revolving round the sun along Venus orbit to know that its absolute dilation T_x compared with T of the earth is 0.0157 second within a year. It is observed from the above mentioned data for absolute dilation that the absolute dilation achieved by human in spacecraft to revolve round the earth, moon and the sun is too little to achieve time travel.

It seems that we have to look for other planet in outer space. It is observed from celestial observation that the largest star in mass discovered is R136a1, its mass is 320 times that of the sun, but its diameter is 3600 times that of the sun, with higher temperature, it is observed from formula (6) that the maximum velocity achieved by flight revolving round it is far below that achieved by revolving round the sun (M/R) is small; the star A₁ is in the second place, its mass is 150 times that of the sun, its diameter is 114 times that of the sun (Reference document 4), assume that there is human being existing in adjacent planet similar to the human being on earth and the spacecraft is to make a fight round the surface of the star, it is observed from calculation that the velocity of the spacecraft is up to 501km/s and that the absolute dilation achieved is 44 seconds within a year, namely the human life is prolonged by 1.4 millionth. It is obvious that the flight can not achieve time travel at all.

It is certain that flight around the black hole shall greatly enhance the absolute velocity and absolute dilation of the spacecraft, however, according to black hole theory, the mass of the black hole similar to football court in size is equal to the mass of four suns. (Reference document 5). It is observed from calculation that the mass of the black hole similar to a grain in size is more than the mass of six trillion people on earth. (Per capita mass is 50kg). If black hole is actual and composed of the known 118 types of elements and then even if all atoms in black hole are collapsed and all electrons around atoms fall on atomic nucleus and turn into neutrons and that the neutrons are compacted together, its volume is one million times more than the volume defined according to black hole theory, indicating that black hole composed of atoms is impossible. We only discuss actual substance and actual celestial body instead of the unproven black hole not composed of atoms, for the vantage, human life is limited and it is impossible for human to reach to black hole.

In summary, the absolute dilation achieved by human in spacecraft to make a flight around celestial body in outer space shall be no more than 44 seconds every year, accordingly, it is impossible to satisfy the basic requirement for human to prolong their life by several times by time travel, consequently, it is observed that it is impossible for human to achieve time travel.

It should be pointed out that J·C·Hafele and R·E·Keating have spent much energy and resources on experiment to learn about the relation between motion velocity and time dilation, their experimental analysis has substantially indicated that the absolute dilation is only related to absolute velocity and unrelated to relative velocity, as a consequence, the discovery of absolute velocity and absolute dilation thanks to J·C·Hafele and R·E·Keating, the paper presents supplementary and induction to their experimental analysis.

References:

1. Wu Shouhuang, *Basics of Theory of Relativity* (1987), Shaanxi Science & Technology Press

2. Fudan University, *Physics* (1985), China Higher Education Press

3. Internet-Baidu.com (2014) Time Travel

4. Internet-Baidu.com (2014) Cosmic Planet

5. Internet-Baidu.com (2014) Black Hole